Eyes on the Sky

# Uranus

by Linda George

THOMSON
GALE™

San Diego • Detroit • New York • San Francisco • Cleveland
New Haven, Conn. • Waterville, Maine • London • Munich

THOMSON
GALE

Cover photo: © NASA/Dr. James Elliot/Photo Researchers
© Julian Baum/SPL/Photo Researchers, 14
© CORBIS, 24
© Corel Corporation, 19, 21, 35, 36
© Araldo de Luca/CORBIS, 9 (right)
Jeff DiMatteo, 6, 17, 28
© Mark Garlick/SPL/Photo Researchers, 29
© A. Gragera, Latin Stock/Science Photo Library/Photo Researchers, 23
NASA, 13
© NASA/Photo Researchers, 5
© NASA/Roger Ressmeyer/CORBIS, 26, 31, 39 (left and right)
© NASA/SPL/Photo Researchers, 33
© Robin Scagell/SPL/Photo Researchers, 40
Martha Schierholz, 10
© Space Telescope Science Institute/NASA/Science Photo Library/Photo Researchers, 16
© SPL/Photo Researchers, 8
© Victoria & Albert Museum, London/Art Resource, 9 (left)



*For more information, contact*
KidHaven Press
27500 Drake Rd.
Farmington Hills, MI 48331-3535
Or you can visit our Internet site at http://www.gale.com



LIBRARY OF CONGRESS CATALOGING-IN-PUBLICATION DATA

George, Linda.
Uranus / by Linda George.
p. cm.—(Eyes on the sky)
Includes bibliographical references and index.
Summary: Discusses the discovery of Uranus and characteristics of the planet, its rings, and its moons.
ISBN 0-7377-1003-9 (hardback : alk. paper)
1. Uranus (Planet)—Juvenile literature. [1. Uranus (Planet)]
I. Title. II. Series.
QB681 .G46 2003
523 .47—dc21

2002009876

Printed in the United States of America

# Table of Contents

# 1 Discovering the Blue-Green Planet

Uranus is a planet in our **solar system**. The solar system is made up of nine known planets that **orbit**, or circle, the sun. Uranus is the seventh planet from the sun and the seventh planet to be discovered. The planet lies 1.736 billion miles from the sun.

Uranus is shrouded in gaseous clouds, making it one of the "gas giants," along with Jupiter, Saturn, and Neptune. Gas giants have no solid outer surface, but they are believed to have rocky cores. Scientists believe that the core of Uranus might be made of rock and ice. They also estimate that the core is about nine thousand miles in diameter—approximately the size of Earth.

It is impossible to know for sure if Uranus has a solid core because of the thick layers of gas and superheated water in the planet's atmosphere. No telescope can penetrate these dense layers to find out if anything lies at the heart of Uranus.

## Size

Uranus is the smallest of the other three gas giants. Jupiter is about twenty-two times larger than Uranus. Saturn is almost seven times larger than Uranus, and Neptune is just slightly larger than Uranus.

Even though Uranus is the smallest gas giant, it is still much larger than Earth. The diameter, or distance across the center, of Uranus is 31,763 miles. This means its diameter is roughly four times that of Earth. About sixty-three Earths would fit inside Uranus.

Uranus is a gas giant whose surface is made up of gases and water.

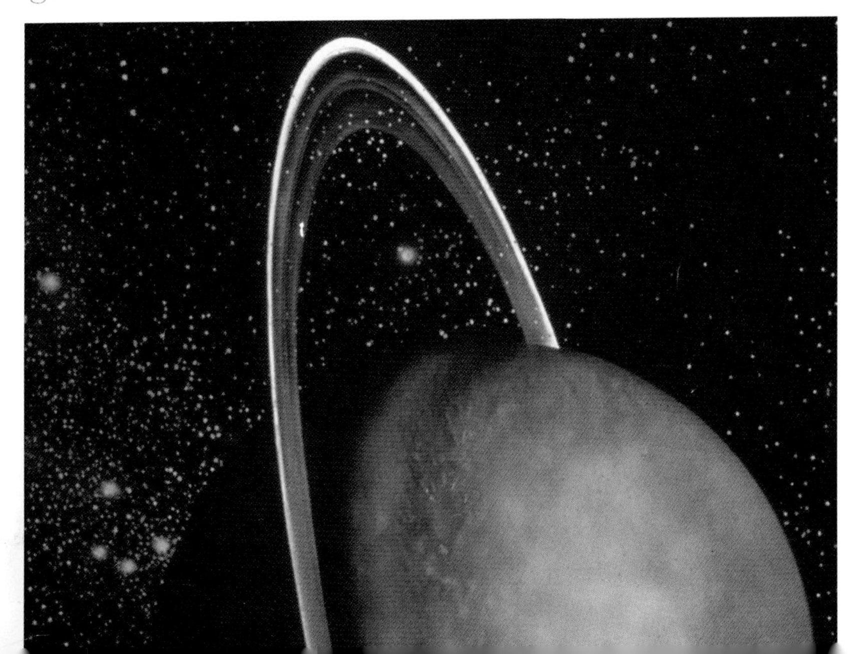

Because Uranus is so much larger than Earth, it is reasonable to expect **gravity** to be greater on Uranus than on Earth. But, it is not. Uranus's gravity is actually slightly less than Earth's gravity. This is because there is less solid material, or mass, on Uranus than on Earth. Planets with higher mass have more gravity than those with less mass. Because of Uranus's lesser gravity, if a child weighed eighty pounds on Earth, that child would weigh only seventy-two pounds on Uranus.

## Atmosphere

The atmosphere of Uranus is much different from that of Earth. On Earth, the atmosphere is

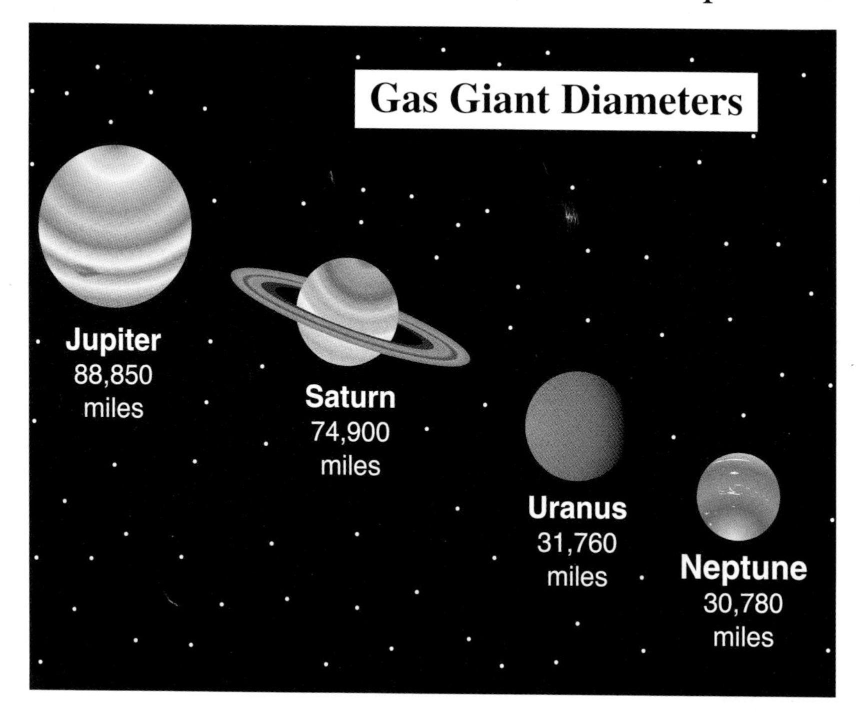

made of mostly oxygen and nitrogen. These gases are not poisonous for human beings to breathe. Uranus's cloudy atmosphere is made mostly of **hydrogen** (83 percent), with lesser amounts of **helium** (15 percent) and **methane** (2 percent) and small amounts of **ammonia** and water. A human being breathing Uranus's atmosphere would quickly choke to death.

The contents of the atmosphere determine the color of a planet. From space, Earth appears blue. Through a telescope, Uranus and Neptune appear bluish green. Jupiter and Saturn appear yellow, orange, and brown. The methane in Uranus's atmosphere absorbs red light and reflects blue and green light, creating its blue-green color. When it was discovered in 1781, the color of Uranus was one of the few things that could be determined about the planet's surface.

## Discovery

William Herschel, an amateur astronomer, discovered Uranus on March 13, 1781, with a telescope he had built in his basement workshop in Bath, England. The telescope was seven feet long and had a lens that was six and a half inches in diameter. It was better than other telescopes of that time. Herschel was observing the **constellation** Gemini when he spied a blue-green disk.

Herschel thought the disk was a new comet. But then he realized the new heavenly body did

William Herschel carefully observes the heavens as his sister, Caroline, records his findings.

not behave like a comet. Its orbit around the sun was almost circular, instead of long and oval-shaped like a comet's orbit. The blue-green disk had to be a planet, twice as far from the sun as Saturn, the farthest planet known at that time.

## Naming the Planet

News of Herschel's discovery quickly reached King George III of England. Herschel wanted to name the new planet George's Star, after the king, but other astronomers disagreed. The planets had traditionally been named after Greek and Roman gods. In Roman mythology, Jupiter was the father of Mars, and Saturn was the father of Jupiter. To continue this tradition, the new planet was named Uranus, in honor of the mythical father of Saturn.

Even though the planet was not named after the king, George III rewarded Herschel for his historic discovery by naming him the King's Astronomer.

Roman gods Mars (left) and Jupiter (right) were the namesakes for two planets. Uranus was also named after a Roman god.

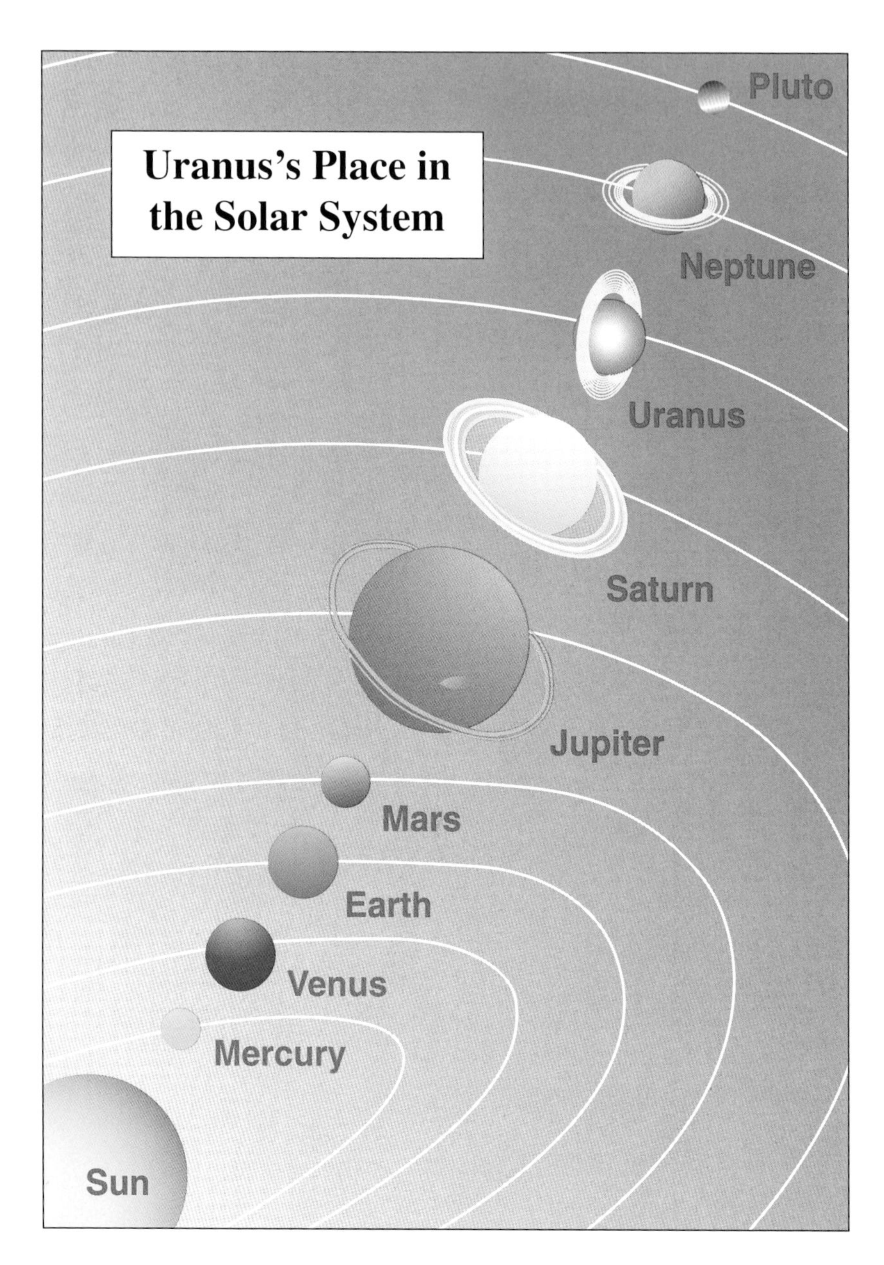


## More Discoveries

Herschel kept watching Uranus and made more discoveries about the planet. He determined that Uranus traveled around the sun every eighty-four years. Then, six years after discovering the planet, he identified two of its satellites, or moons. Astronomers named them Oberon and Titania.

There was much more to learn about Uranus. But because the planet was so far from Earth, the information was slow in coming. Even as telescopes improved over the centuries, Uranus was so far away that little could be revealed by observing it from Earth.

The only way to learn anything in detail was to send a spacecraft to the planet. That did not happen until two hundred years after Herschel's discovery. When NASA spacecraft *Voyager 2* visited the planet in 1986, it confirmed a lot of information about the planet. It also made new discoveries that surprised and delighted scientists.

# 2 Visiting the Distant Planet

Most of what scientists know about Uranus came from information transmitted back to Earth by the *Voyager 2* spacecraft. In the summer of 1964 officials at NASA planned an unmanned voyage to Jupiter and Saturn. A team of scientists was assembled to construct a spacecraft and program it for exploration of these two gas giants.

Gary Flandro, a doctoral student at NASA's Jet Propulsion Laboratory in Pasadena, California, was asked to join the group of scientists planning the mission. He noted that in 1980, all the outer planets except for Pluto would be lined up on the same side of the sun. This alignment happens only once every 180 years. A spacecraft launched from Earth in 1977 or 1978 could travel past Jupiter and Saturn and move on to the outer planets, Uranus and Neptune.

NASA officials were delighted with the possibility of extending *Voyager*'s journey. The project committee quickly announced that *Voyager* would make the "Outer Planets Grand Tour." Thirteen years later, on August 20, 1977, *Voyager 2* blasted off from Cape Canaveral, Florida. *Voyager* did not reach Uranus for eight years. If a child had been in kindergarten when *Voyager* was launched, that child would be in the eighth grade when the spacecraft finally visited Uranus.

A Titan rocket (like the one pictured here) launched *Voyager 2*, the spacecraft that gathered information about Uranus.

## *Voyager*'s Journey

On January 24, 1986, after having traveled nearly 2 billion miles from Earth, *Voyager 2* passed within 50,700 miles of Uranus, less than 10 miles from where scientists had planned for it to be. Experts compared this amazing accuracy to making a hole in one with a golf ball from a thousand miles away.

*Voyager* had only about six hours to get information about the planet before continuing on its journey to Neptune. Still, the spacecraft sent back more information than scientists and astronomers ever expected.

*Voyager 2* passes by Uranus and transmits information about the planet back to Earth.

In many ways, the information *Voyager 2* collected about Uranus was surprising. Compared to the other planets, and based on what William Herschel and other astronomers had learned about Uranus since its discovery, the blue-green planet was unique in many ways.

## Rotation

Perhaps the most startling fact that *Voyager* revealed was that Uranus lies on its side. Earth and the rest of the planets in the solar system **rotate** primarily side to side in their orbits, like a top spinning on its point. Most planets tilt a little to one side as they rotate through space. For example, Earth tilts 23.45 degrees. Uranus, however, is tilted a remarkable 98 degrees. This means Uranus is tipped over on its side, and the top of the planet—the geographic North Pole—points slightly downward. As a result, Uranus rotates from top to bottom, like a ball rolling across the floor.

No ones knows why Uranus is tipped sideways, but scientists believe a huge body from space, perhaps as large as Earth, crashed into Uranus millions of years ago and knocked it on its side. Because Uranus lies on its side, the planet has especially long days, nights, and seasons.

Earth spins on its **axis**—an imaginary line that runs down the center of the planet—once

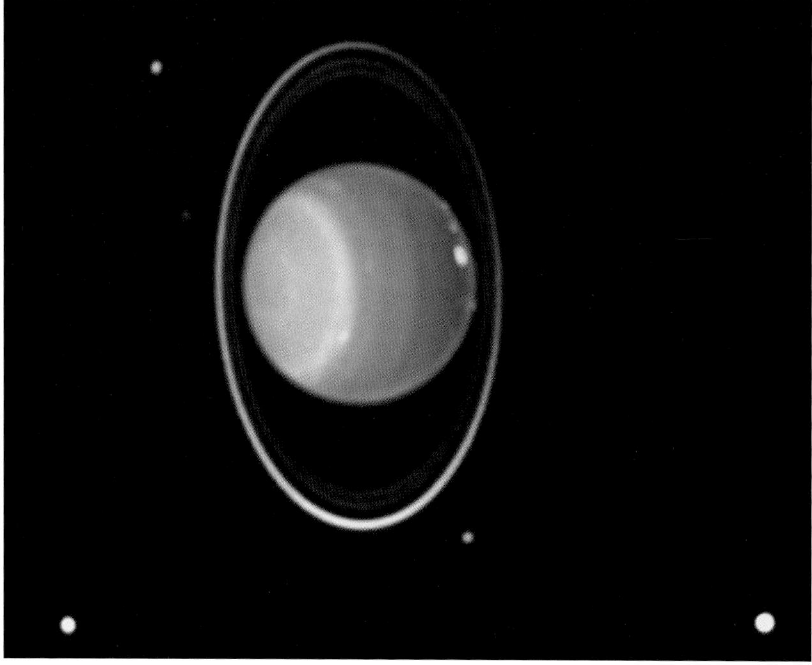

Uranus lies on its side due to the sharp tilt of its axis.

every twenty-four hours. This is called an Earth day. Even though Uranus is much larger, it spins faster. *Voyager*'s readings helped scientists determine that a day on Uranus is only seventeen hours and fourteen minutes in Earth time.

While it takes Earth about 365 days, or one year, to travel around the sun, it takes about eighty-four Earth years for Uranus to travel around the sun one time. This is because Uranus is so much farther from the sun than Earth is. Earth will orbit the sun eighty-four times while Uranus orbits the sun only once.

While orbiting the sun, planets experience seasons. Unlike Earth, where seasons are roughly equal in length, the North Pole of Uranus faces the sun for half the year, while the South

Pole is dark. Then, the southern half is sunny while the northern half is dark. This happens because Uranus is tipped on its side. When *Voyager* visited the planet, the South Pole was facing the sun, and the North Pole was dark.

On Uranus summer and winter each last forty-two Earth years because of its unusual 98-degree tilt. Uranus's weather during these long seasons is not like the weather on Earth, where temperatures sometimes change drastically. Uranus remains the same temperature all over the planet, whether on the sunny side or on the dark side. The constant temperature results from

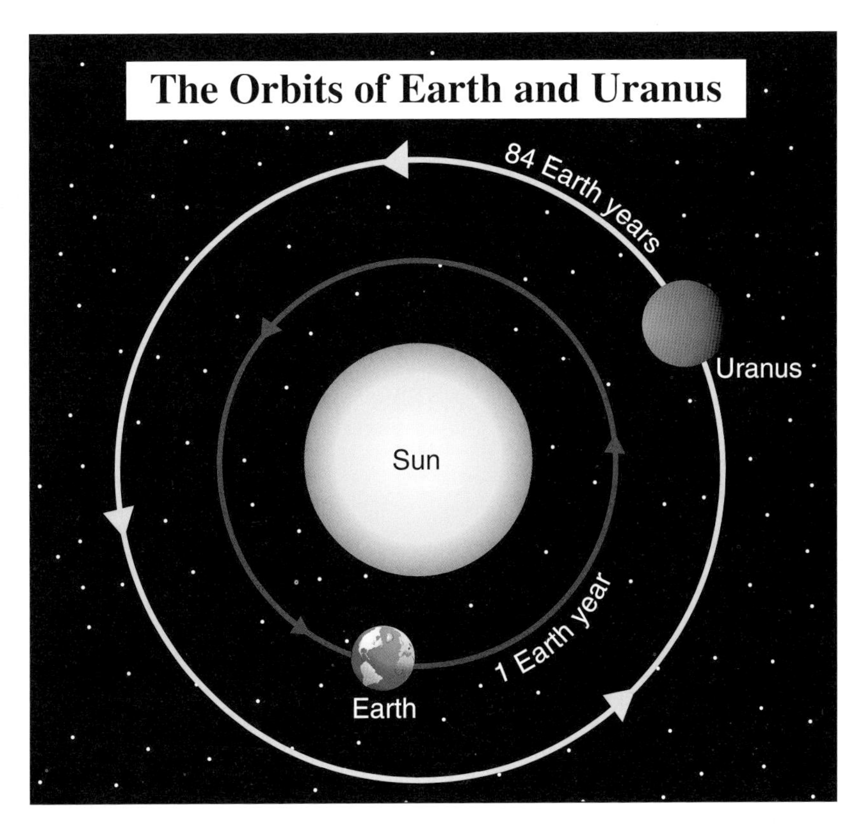

the fact that the planet's core does not radiate heat it receives from the sun back to the atmosphere. This unusual trait may also explain why there are never storms in any season on Uranus.

## Strong Winds but No Storms

Unlike the other gas giants, Uranus never has surface storms such as hurricanes or cyclones. This does not mean that the wind does not blow. *Voyager* measured the winds halfway between the poles and the equator and discovered that the wind blows between ninety and 360 miles per second on the surface of the planet. These winds blow in the direction the planet rotates.

At the equator, though, *Voyager* measured winds blowing about 225 miles per second in the opposite direction from winds farther north and south. Winds blowing in different directions on Earth usually cause storms. This does not happen on Uranus, but no one can explain why.

## Surface of the Planet

As *Voyager* passed by Uranus, it was also able to take the planet's temperature. The temperature of the atmosphere on the surface of the planet is extremely cold, about minus 365 degrees Fahrenheit. But, in contrast, the temperature of the core of the planet is extremely

Uranus's swirling cloud layer presses down on the surface of the planet, creating extreme heat in the core.

hot. The core reaches temperatures as high as 14,500 degrees Fahrenheit. The tremendous pressure of the planet's heavy cloud layers causes this intense heat. The cloud layers press down on the core, causing it to get hot. This pressure is about 8 million times greater than the atmospheric pressure on Earth.

When water on Earth gets hot, it boils at 212 degrees Fahrenheit. But the water inside Uranus never boils. Because of the tremendous heat and pressure, the water becomes electrically charged, or **ionized**, but it does not boil. Instead, the particles of ionized water cause a magnetic field to form around the planet.

## Magnetic Field

Every planet has a magnetic field that surrounds it like a bubble. This magnetic field is called a magnetosphere. The magnetosphere of Uranus is about fifty times greater than Earth's magnetic field.

On Earth a compass needle points north, but it does not point toward the geographic North Pole—the northern tip of Earth's axis. Instead, it points toward Earth's magnetic North Pole, which is currently about 12 degrees south of the geographic North Pole.

On Uranus, the magnetic North Pole is much farther away from the geographic North Pole. It lies at a specific point, 60 degrees away from the geographic North Pole. If this were true on Earth, and Earth's magnetic north were extended 60 degrees from geographic north, a compass needle would point toward an area about one hundred miles west of San Antonio, Texas.

Another magnetic difference between Earth and Uranus involves the "tails" of their magnetospheres. Electrically charged particles from the sun stream around all the planets in the solar system. This stream of particles is called solar wind. Solar wind causes particles in the

Uranus's atmosphere radiates as the solar wind billows above the planet's surface.

magnetic field to stream out behind a planet, like the tail of a comet.

Behind most planets, the magnetic tail streams in a straight path. But Uranus is different. Uranus's tail twists and spirals like a corkscrew. This happens because of Uranus's unusual rotation.

## *Voyager*'s Other Discoveries

*Voyager* learned a great deal about Uranus and how it is different from the other gas giants. The spacecraft also identified ten more of Uranus's moons and gathered information about the rings circling the planet. These rings were discovered just before *Voyager* was launched.

# 3 The Rings of Uranus

As scientists continue to study Uranus, the planet rewards them with surprising discoveries. One of the most astounding discoveries, though, was made on March 10, 1977. On that day, astronomers were observing Uranus as it traveled in front of a faint star in the constellation Libra. They expected that when Uranus passed in front of the star, the star would temporarily disappear from sight.

The astronomers were shocked, however, when the star disappeared thirty-five minutes before the planet had lined up with it. The star soon reappeared. But during the period before Uranus was in line with the star, the star "blinked" on and off four more times. Likewise, after Uranus passed by the star, the star disappeared and reappeared five more times. The only explanation was that something on either side of Uranus had also passed in front of the star. Because the star "blinked out" in the same pattern on both sides of Uranus, the answer had to be that the planet has rings.

The discovery that Uranus has rings was important to astronomers. The only other planet at that time known to have rings was Saturn. (Jupiter's rings were not discovered until 1979. Neptune's rings were discovered ten years after that, in 1989.) Scientists eventually concluded that rings are a common feature of planets, especially the gas giants, and not unique to Saturn.

Like the other three gas giants, Jupiter (shown here with several of its moons) also has rings.

## Vertical Rings

Unlike Jupiter, Saturn, and Neptune, whose rings circle the planet more or less horizontally, Uranus's rings circle the planet vertically. This is because Uranus lies on its side, and the rings circle the planet's equator.

From Earth, astronomers identified nine rings around the planet. This was quite an achievement because Uranus's rings are very hard to see. Unlike Saturn's spectacular rings, Uranus's rings are so slender and dark that they are nearly invisible.

Uranus and a ring, as seen from one of its moons.

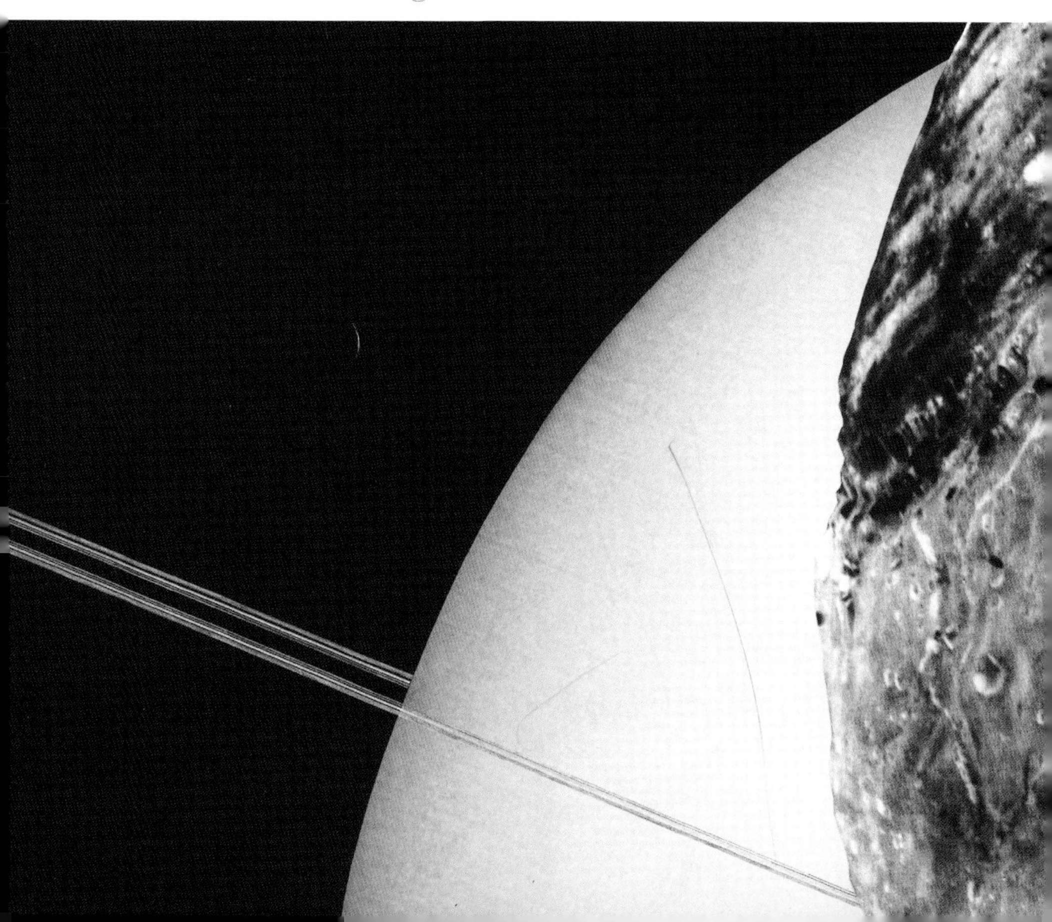

Even with a telescope, astronomers could not see the rings clearly from Earth because they reflect almost no sunlight. Two of Uranus's rings reflect about 60 percent of the sunlight that reaches them. Another, the Epsilon ring, which is farthest from the planet, reflects only 3 percent of the sunlight that falls on it. Two reasons account for the low level of reflectivity of Uranus's rings.

First, the rings are extremely narrow with sharp edges. Most of the rings are less than ten miles wide. The Epsilon ring is the widest. Its width varies from twelve to sixty miles. The rocks and debris that make up the rings are fairly small. Individual chunks of rock range from less than a foot to only about thirty feet in diameter.

Second, the material in the rings is not very reflective. The rings do not contain pieces of ice or large fields of dust, both of which help make Saturn's rings so bright. Instead, the materials in Uranus's rings are naturally dark. Scientists believe the debris in the rings is made of carbon and of frozen methane gas. Carbon is black, and frozen methane is also dark. These dark objects are very difficult to see in the narrow rings.

## Two New Rings

It was not until *Voyager 2* reached Uranus that the planet's rings were observed and analyzed with any accuracy. When *Voyager 2*

A satellite captured this detailed view of Uranus's rings.

journeyed past in 1986, it discovered two additional rings, bringing the total to eleven. Inward from the outer Epsilon ring are the Delta, Gamma, Eta, Beta, and Alpha rings. The rings named 4, 5, and 6 lie inward from there and are barely visible. The two new rings were named 1986U1R, which lies just inside the Epsilon ring, and 1986U2R, which is the closest ring to the planet.

In addition to the eleven identified rings, *Voyager 2* detected evidence of ring fragments. These fragments could have been formed when one of the planet's moons was struck by an asteroid or other object, and then broke into pieces. The presence of ring fragments indicated that the rings are younger than the planet, just as the rings of Jupiter and Saturn are younger than the planets they circle.

Uranus's rings circle the planet at a distance of twenty-six thousand to thirty-two thousand miles from the center of the planet. All of Uranus's rings are distinct, or obviously separate, from each other. However, some scientists believe that the debris in the rings may eventually join and form a single ring around the planet. Others predict the materials may be drawn into the planet by gravity, causing the rings to disappear. It will take millions of years, however, for any such changes to occur.

## Bumping Together

*Voyager 2* determined that the rings closest to Uranus circle the planet at a different speed than the rings farther away. Sometimes, the edges of rings rub or bump up against each other as they spin at their different speeds.

This bumping causes the particles in the rings to bounce around, separate, and move away from the planet. But *Voyager* could see that the Epsilon ring was not getting farther from the planet. Scientists knew something must be holding the ring in place.

When the same spreading process was observed in Saturn's rings, scientists discovered several small moons within the rings. Gravity from these moons stopped the ring debris from floating off into space. Such moons are called **shepherding moons** because they keep the ring material from wandering away.

When *Voyager* passed close to Uranus, it found two moons, Cordelia and Ophelia, on

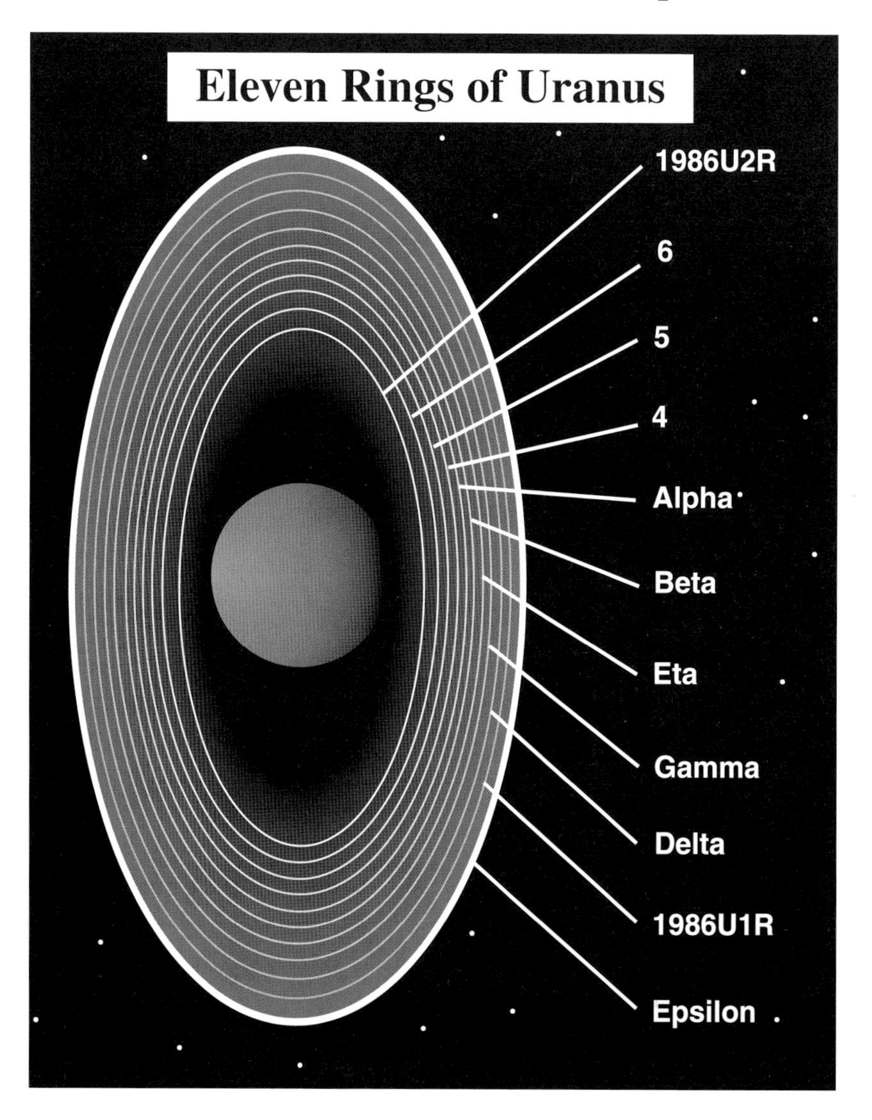

A computer illustration shows Uranus (blue background, left) and five of its largest moons.

either side of the Epsilon ring. The gravity from Cordelia and Ophelia was obviously holding Uranus's rings in place. Astronomers wondered if Uranus had more shepherding moons, but *Voyager* was not able to detect any. Many scientists believed there could be more shepherding moons on the side of Uranus that *Voyager* could not see during its brief observation of the planet.

# 4
# The Moons of Uranus

Uranus has twenty-one known moons. The first five moons were discovered by William Herschel, William Lassell, and Gerald P. Kuiper. *Voyager 2* discovered ten more moons, but it could not see the moons on the far side during the time it visited the planet. Six more moons have been discovered since *Voyager*'s visit to the planet. These moons were discovered using the Hubble space telescope. This telescope orbits Earth and takes pictures of stars and planets and other bodies in space.

## Dark and Hard to See

Uranus's moons consist primarily of water ice and rock, with lesser amounts of carbon and ammonia ice. All but one of the moons contain methane ice. All of these substances are dark, which makes the moons appear gray, brown, or black. As a result, they are dark and difficult to see from Earth. They are also difficult to see because all of them are smaller than Earth's moon.

Astronauts carefully repair the Hubble space telescope. The telescope was used to discover six of Uranus's moons.

Uranus's smallest moons are the shepherding moons, Cordelia and Ophelia. They are only nine miles in diameter, making them hard to locate easily.

The moon Puck is ninety miles in diameter. Although it is the sixth largest moon of Uranus, it is still hard to see. This is because Puck is a very dark moon. Only 7 percent of the sunlight striking Puck is reflected. The rest is absorbed because of the moon's dark color.

The five largest moons are easier to see from Earth. They exhibit characteristics shared by all the moons orbiting Uranus, but each is unique. Their distinguishing features make them some of the most intriguing bodies in the solar system.

## Oberon

William Herschel discovered Uranus's outermost moon, Oberon, six years after he discovered the planet. Oberon is 946 miles in diameter and shaped like a sphere, or ball. It orbits at just over 362,585 miles from the center of the planet.

Oberon's surface is made of dark rock covered with craters caused by collisions with asteroids or other objects from space. Scientists believe these collisions happened long ago. They base their beliefs on studies of the moon. These studies show that little has changed on the surface over the past 3 billion years.

Five large moons orbit Uranus as it glows in the center.

In one of Oberon's ancient craters is a mountain twelve miles high. This mountain is more than twice as high as Mount Everest, the tallest mountain on Earth.

## Titania

Herschel also discovered the moon Titania shortly after he discovered Oberon. Titania is

981 miles in diameter and is also shaped like a sphere. It orbits at 271,162 miles from the center of the planet. Titania has enormous valleys that stretch across its surface. Some of the canyons extend one thousand miles and are up to forty-five miles wide, making them larger than the Grand Canyon on Earth. These winding trenches may have been formed when water inside the moon froze, expanded when the temperature rose, and then cracked the surface of the moon.

Like Oberon, Titania exhibits impact craters from collisions with space debris. Some of these craters have rings that resemble the ripples caused by a rock thrown into a pond.

## Umbriel

The moon Umbriel was discovered in 1851 by William Lassell. It is 726 miles in diameter and spherical in shape. Umbriel orbits at 165,320 miles from the center of the planet.

Umbriel is the darkest of Uranus's moons. This is because there has been a lot of volcanic activity on the moon. This volcanic activity coated the surface with materials containing black carbon. In some places this carbon covering is six miles deep. Unlike the surface material on Uranus's other moons, the surface of Umbriel does not show signs of methane. Scientists

Some canyons on the moon Titania are larger than Earth's Grand Canyon (pictured).

cannot explain why Umbriel lacks this common compound.

Umbriel's surface resembles the bumpy, dimpled skin of an orange. It does not have the cliffs, valleys, ice floes, or twisting faults found on other moons.

A single white ring exists on the surface in a crater named Wunda. The ring was called the "fluorescent Cheerio" by scientists observing the

Umbriel's surface looks a little like this landscape in Hawaii.

photographs sent back by *Voyager 2.* The ring probably consists of ice from water that bubbled to the surface from the moon's interior after some volcanic eruptions.

## Ariel

William Lassell also discovered the moon Ariel, soon after Umbriel's discovery. Ariel is 720 miles in diameter and shaped like a lumpy sphere. The moon orbits at 118,707 miles from the center of the planet.

Ariel has smooth-floored valleys, similar to some of the valleys on Earth. Ariel also has craters, many of which are as much as six miles in diameter. The surface seems to have cracked, forming valleys that were filled in by volcanic material. Ariel has several bright patches in some of the younger craters. Scientists do not know what these patches are, but some believe they may be ice.

## Miranda

Miranda, innermost of the five largest Uranian moons, has been called "bizarre" by scientists working on the *Voyager* mission. Discovered in 1948 by Gerald P. Kuiper, Miranda is 290 miles in diameter and is shaped like a lumpy sphere. Miranda's orbit is not circular. The moon travels in an elliptical—or slightly flattened circular—

orbit. Thus, the distance the moon is from the center of Uranus varies from 78,521 to 82,878 miles.

The surface of this moon consists of canyons, ridges, and high cliffs. Some of the jagged cliffs circle craters and resemble the points of a crown. For this reason they are called coronae, from the Latin word for "crowns." These coronae were probably formed by volcanic eruptions that spewed icy material onto the surface from the interior of the moon. The largest crowns were named Arden, Inverness, and Elsinore, after the settings of some of English playwright William Shakespeare's plays.

## Coronae Characteristics

The Arden corona is also called "the racetrack." In the basin of the oval-shaped crater are grooves lying side by side, resembling the lanes of a racetrack. The ridges on either side of the racetrack are so high, a rock dropped from the top of one would take ten minutes to fall to the ground, twelve miles below.

The edges of the Inverness corona have sharp peaks, while the Elsinore corona is smoother. The smoothness may be caused by repeated eruptions of icy lava from the center of the moon that eventually flattened the ring of sharp peaks.

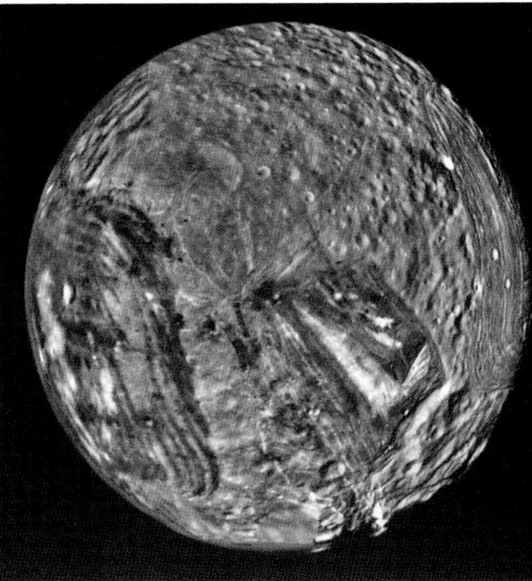

Ariel (left) has a smooth surface while Miranda (above) features jagged cliffs and high mountains.

Miranda has different landscapes not found together on any other body in the solar system. Scientists have come to the conclusion that Miranda must have been struck by a huge body traveling through space. They think the moon must have shattered into chunks and pieces. These pieces, over thousands of years, were pulled back together by Miranda's gravity, but not into their original positions. Instead, the chunks formed a moon with smooth areas next to towering mountains and deep valleys,

An astronomy buff peers through a telescope to take a closer look at outer space.

and old, weathered portions next to newer, ridged sections.

## Only the Beginning

Uranus's moons display some of the most unusual features of any heavenly bodies observed by scientists through telescopes from Earth, from *Voyager 2,* or from the Hubble space telescope. In time, astronomers hope to learn more about the moons, rings, and Uranus itself, by studying all the information they have gathered.

**ammonia:** A compound formed when nitrogen and hydrogen combine. Ammonia is one of the gases found in Uranus's atmosphere.

**axis:** An imaginary line through the center of a planet; the planet spins, or rotates, on its axis.

**constellation:** A collection of stars.

**gravity:** The force of attraction. On a celestial body gravity keeps matter from floating off into space.

**helium:** A light, colorless gas that does not burn. Helium makes up a large part of the atmosphere of Uranus.

**hydrogen:** A light, colorless gas that catches fire easily. Hydrogen is one of the primary elements in the atmosphere of Uranus.

**ionized:** Electrically charged.

**methane:** A colorless gas formed when carbon and hydrogen combine. Methane is another of the gases found in Uranus's atmosphere.

**orbit:** The invisible path of one heavenly body circling another. Uranus's moons orbit the planet.

**rotate:** To spin on an axis. Uranus rotates once every seventeen hours, fourteen minutes.

**shepherding moons:** Moons on either side of a planet's rings that keep the matter in the rings from drifting into space.

**solar system:** The sun, as well as the planets, moons, asteroids, and comets that orbit the sun.

***Voyager 2:*** The space probe launched in 1977 that was sent to gather information and take pictures of Jupiter, Saturn, Uranus, and Neptune. *Voyager 2* visited Uranus in 1986.

## Books

Larry Dane Brimner, *Uranus*. Danbury, CT: Childrens Press, 1999. Simply written, up-to-date information about Uranus. Captioned color photographs, glossary, and websites are included.

Dennis B. Fradin, *Uranus*. Danbury, CT: Childrens Press, 1999. Discusses how Uranus was named and what astronomers have learned about the planet.

Robin Kerrod, *Uranus, Neptune, and Pluto*. Minneapolis: Lerner, 2000. Describes the characteristics of these three planets. Includes computer-generated photographs of the planets and information about the *Voyager* mission.

Melissa Stewart, *Uranus*. Danbury, CT: Franklin Watts, 2001. Describes the discovery, explorations, atmosphere, orbit, and moons of Uranus.

Luke Thompson, *Uranus*. New York: Rosen, 2000. Examines the history, unique features, and exploration of Uranus.

Gregory L. Vogt, *Uranus*. Brookfield, CT: Millbrook Press, 1994. Impressive photographs and images taken from Earth, from *Voyager 2*, and from the Hubble space telescope.

## Websites

**National Space Science Data Center** (http://nssdc.gsfc.nasa.gov). A section of this website contains facts about Uranus, comparisons with all the planets in the solar system, and information about rings and moons. Includes links to a Uranus fact sheet and information about the *Voyager 2* mission.

**Solar System Exploration** (http://sse.jpl.nasa.gov). An excellent, user-friendly site with general information, statistics, and images. Links to other sites about Uranus and the solar system.

**Starchild** (http://starchild.gsfc.nasa.gov). This excellent site for young astronomers includes sound, music, and graphics. There are two levels of information, a full glossary of terms, and links to other sites to learn about astronomy.

**Students for the Exploration and Development of Space** (www.seds.org). The planets section of this website has good general information about Uranus. The site is easy to understand and has a list of moons and some basic information about them in chart form. Also contains links to information about *Voyager 2*, the other gas giants, William Herschel, the Hubble space telescope, and other Uranus sites.

**Views of the Solar System** (www.solarviews.com). A multimedia presentation of information on Uranus. Includes extensive information about the moons and rings and lots of color photographs. It also has tables of specific data about the planet and its moons and rings. Three animations can be downloaded: the Discovery of Uranus, the Magnetic Field, and Core and Magnetic Field.

**Welcome to the Planets** (http://pds.jpl.nasa.gov). This site has technical data about the planets. Includes numerous color photos, some incredibly detailed, of Uranus's moons and rings.

# Index

# About the Author

Linda George is the author of more than thirty nonfiction books for children and teens. She lives in the New Mexico mountains with her husband, Charles, who also writes for KidHaven Press. Together they wrote a history of Texas for KidHaven's Seeds of a Nation series.